The INSIDE GUIDE

STEM CAREERS

Jobs in Environmental Science

By Beatrice Harris

Cavendish Square

Published in 2024 by Cavendish Square Publishing, LLC
2544 Clinton Street Buffalo, NY 14224

First Edition

Cataloging-in-Publication Data

Names: Harris, Beatrice.
Title: Jobs in environmental science / Beatrice Harris.
Description: Buffalo, New York : Cavendish Square Publishing, 2024. | Series: Inside guide: STEM careers | Includes glossary and index.
Identifiers: ISBN 9781502670304 (pbk.) | ISBN 9781502670311 (library bound) | ISBN 9781502670328 (ebook)
Subjects: LCSH: Environmental sciences–Vocational guidance–Juvenile literature. | Environmental sciences–Juvenile literature.
Classification: LCC GE60.H36 2024 | DDC 577.27–dc23

Editor: Therese Shea
Copyeditor: Shannon Harts
Designer: Deanna Lepovich

The photographs in this book are used by permission and through the courtesy of: Cover DC Studio/Shutterstock.com; p. 4 Paul shuang/Shutterstock.com; p. 6 Moh. Saefudin/Shutterstock.com; p. 7 Pornpimon Ainkaew/Shutterstock.com; p. 8 (main) Viewfoto studio/Shutterstock.com; p. 8 (sidebar) Seal of the United States Environmental Protection Agency/Wikimedia Commons; p. 9 KT Stock photos/Shutterstock.com; p. 10 Actium/Shutterstock.com; p. 12 Pixel B/Shutterstock.com; p. 13 APChanel/Shutterstock.com; p. 14 M2020/Shutterstock.com; p. 15 Dennis W Donohue/Shutterstock.com; p. 16 PeopleImages.com - Yuri A/Shutterstock.com; p. 18 TasfotoNL/Shutterstock.com; p. 19 Lena_viridis/Shutterstock.com; p. 20 (main) Sundry Photography/Shutterstock.com; p. 20 (sidebar) lev radin/Shutterstock.com; p. 21 Viktor Birkus/Shutterstock.com; p. 22 smite2016/Shutterstock.com; p. 24 (top) Roman Mikhailiuk/Shutterstock.com; p. 24 (bottom) vanhurck/Shutterstock.com; p. 25 (main) NASA/Karl Shreeves; p. 25 (inset) Edie thumb 33 ft/Wikimedia Commons; p. 26 Drazen Zigic/Shutterstock.com; p. 27 Tongpool Piasupun/Shutterstock.com; p. 29 (left) goodluz/Shutterstock.com; p. 29 (right) Riderfoot/Shutterstock.com.

CPSIA compliance information: Batch #CWCSQ24: For further information contact Cavendish Square Publishing LLC at 1-877-980-4450.

Printed in the United States of America

CONTENTS

The natural world, or environment, needs protecting. Scientists can help!

Chapter One

WORKING FOR THE ENVIRONMENT

If you keep up with the news today, you've probably heard terms like "climate change emergency," "environmental disaster," and "global warming." It's easy to feel discouraged about what one person can do about the state of our planet, but many people have been inspired to take action. Some protest to persuade governments and businesses to alter harmful policies. Some are changing their daily habits, adopting a more Earth-friendly lifestyle. Others are devoting their lives to studying and taking care of the environment. These people who use their knowledge, interests, and skills to safeguard the natural environment—including the organisms in it—are called environmental scientists.

Fast Fact

What is the environment, exactly? It's all the conditions that surround living things and can affect their growth, health, and survival.

So Much Science!

Earth doesn't have just one kind of environment. Mountains, deserts, grasslands, forests, swamps, oceans, lakes, rivers ... these are all **habitats** that provide life-forms with what they need to live. Environmental scientists seek to protect these places, the organisms in them, and the air and water the organisms need to survive. It's a huge task!

A group of people in Indonesia plant mangrove trees to protect coastal areas from erosion.

So, most environmental scientists specialize, or become experts in, a smaller area of study. In fact, environmental science covers many kinds of sciences, including

- ecology (the study of how organisms interact with their environment);
- geology (the study of the history of Earth, especially as it has to do with rocks);
- meteorology (the study of Earth's atmosphere and its weather);
- biology (the study of organisms and their life processes);
- chemistry (the study of substances and their properties);
- engineering (the application of science and math to the design and building of machines, structures, and processes); and
- physics (the study of the properties of matter and energy).

Even within these sciences, there are more focused fields of study.

An environmental scientist takes samples of river water to find out if it's polluted.

Fast Fact

According to the U.S. Bureau of Labor Statistics, the median salary for people with the job title "environmental scientist" or "environmental specialist" was $76,530 in 2021. (Median means half of the workers earned more than that amount and half earned less.)

A Growing Field

As people around the world are recognizing the perils of climate change and pollution, the demand for environmental scientists has grown. Most of the careers that environmental scientists pursue require a college degree, but not all. Some may require more education after college and often require experience too.

THE EPA'S ENVIRONMENT

The U.S. Environmental Protection Agency (EPA) was created to protect the air, land, and water of the United States. The EPA has a detailed explanation of what the U.S. environment is: "(A) the navigable waters, the waters of the **contiguous** zone, and the ocean waters of which the natural resources are under the exclusive management authority of the United States under the Fishery Conservation and Management Act of 1976, and (B) any other surface water, groundwater, drinking water supply, land surface or subsurface strata [layer], or ambient [surrounding] air within the United States or under the **jurisdiction** of the United States."

The EPA was created in 1970 to oversee national environmental efforts, such as the Clean Air Act. Today, it employs over 15,000 people, including environmental scientists.

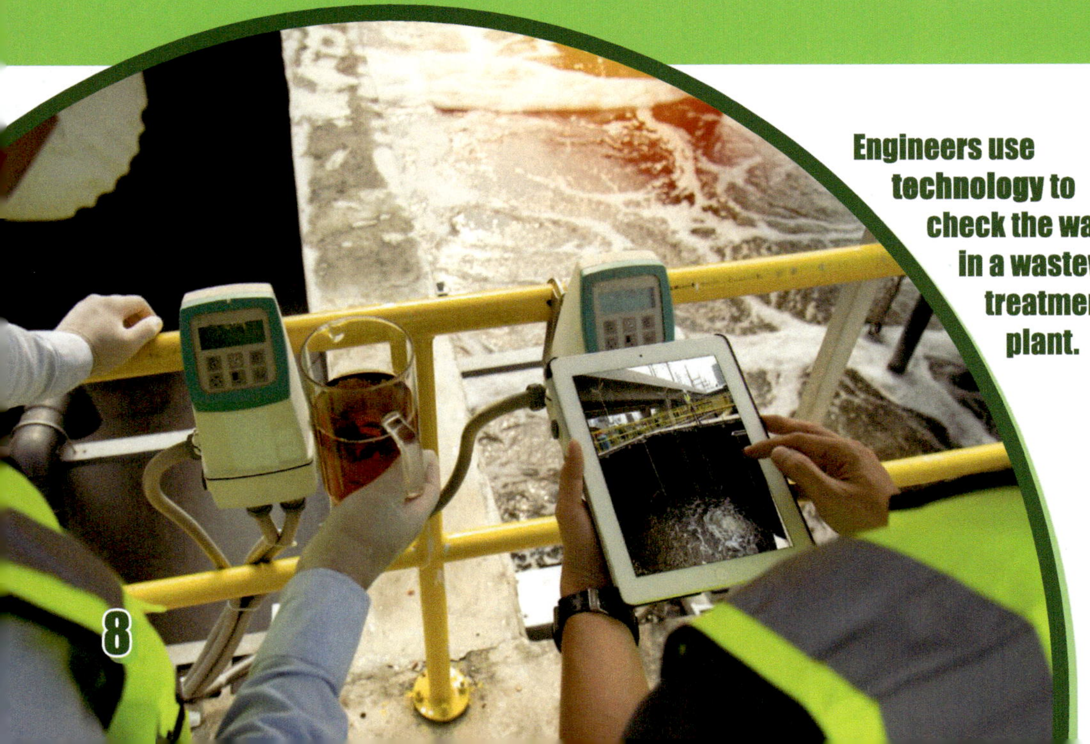

Engineers use technology to check the water in a wastewater treatment plant.

As you read about these careers, think about what you'd like to specialize in and whether you'd want to work in a lab, at a computer, or "in the field." This phrase means doing work in nature, such as at sea, in a forest, or, yes, in an actual field! Do you want to work with nearly extinct animals or really odd plants? Do you want to study high mountains or the deep ocean floor? If you have an interest in nature, a job in environmental science is out there that could help you make a difference.

Fast Fact

Some people who study environmental science in college go on to become lawyers. They use the law to fight for important environmental causes.

Botanists are scientists who study plants. Many plant species are greatly affected by even a small amount of pollution.

Air pollution analysts may also be called air quality technicians.

Chapter Two

AIR AND WATER

The quality of the air and water around us is very important! Not only does clean air smell better, but it also can affect our health. Breathing in polluted air can cause major problems for people with **asthma** and even affect the lungs and heart of people who don't. Polluted water is both unpleasant to look at and poses risks for the organisms that live in it or near it. This includes people. Water can transmit serious diseases too.

Some environmental scientists focus their work on one or both of these important areas. They help locate problems. They may suggest and apply solutions too.

Air Pollution Analyst

When you spot smoke from a wildfire or factory smokestack, it's not a surprise that the air quality in that area is poor. However, air quality isn't always something you can see. Air pollution analysts are scientists who know how to collect samples of air, measure the amount of pollutants in it, and determine how safe the air is for people to breathe.

Fast Fact

The EPA has 5,000 air quality monitors collecting data around the United States and U.S. territories.

This is an example of an air quality monitoring station.

Fast Fact

Scientists called air quality engineers may be out in the field, working on air pollution cleanup more than analysts. However, job descriptions differ.

The analysts may be out in the field, setting up equipment called air quality monitors, or in a lab analyzing samples. Sometimes they do both. Analysts need to know where to place equipment, taking into account weather conditions such as wind direction. This equipment can often both measure the air quality and remotely transmit the data to scientists. If air quality is considered poor, the air pollution analyst works with others to inform the public and suggest a response to keep people safe and even reduce the amount of pollution.

Air quality analysts often need a college degree. They need to have knowledge of the environment and have experience with electronics and machinery in order to operate equipment. They may work for a government or for a business concerned about its environmental impact.

Water Quality Scientist

Water quality scientists, sometimes called water pollution scientists, often focus on testing and analyzing drinking water, water under the ground, or surface water. Depending on where they work, they may both collect

Human activities can have unintended effects on water supplies. The need for water quality scientists is expected to increase in the coming years.

RENEWABLE RESOURCES ENGINEER

The use of fossil fuels—coal, oil, and natural gas—to produce power is a problem. When burned, they release greenhouse gases, such as carbon dioxide, which trap heat in Earth's atmosphere, contributing to climate change. Renewable resources engineers work to design, build, and improve the machines and processes that use renewable resources, or resources that are renewed by natural processes so they don't run out. Many people are hopeful about the future because of technology that uses renewable resources, such as sunlight, wind, and water, to create energy. Renewable resources engineers usually study traditional engineering fields, such as mechanical, chemical, and electrical engineering. Then, they work for renewable energy businesses and organizations.

Solar and wind farms require renewable resources engineers to build and maintain equipment.

Many U.S. bodies of water, including New York's Hudson River, shown here, are polluted. The EPA works on cleanup efforts.

samples and study them in a lab or just do the lab analysis.

These scientists may also have to figure out why water quality has changed. Not all sources of water pollution are obvious. Sometimes floods can carry pollutants a long way, for example. Water quality scientists may need to understand how certain water systems, such as a city's wastewater pipes, work.

These scientists usually have a college degree. They should have knowledge of chemistry too. Many water quality scientists work for businesses or the government.

Fast Fact

According to a 2021 United Nations environmental study of more than 75,000 bodies of water in 89 countries, more than 40 percent were considered very polluted.

Some environmental scientists study plants, which are an important part of habitats.

Chapter Three

ON THE WILD SIDE

Do you enjoy learning about interesting plants, such as the Venus flytrap that traps and digests insects? Maybe you're more of an animal lover who enjoys bird-watching. Some environmental scientists pursue their careers because of a passion for plants and animals. They found jobs in which they pair an interest in the environment with the organisms that live in it. Even within these careers, they specialize in the areas that attract them most.

Wildlife Biologist

Wildlife biologists work to protect wildlife. They may be concerned with a certain animal or with a variety of animals in a certain area. Their tasks may include estimating the population of a kind of animal in a habitat to see if the population is healthy. Based on that number, they might work with others to decide on hunting or fishing rules. They might discover that a species is endangered, or in danger of becoming extinct, and work with authorities on passing laws to protect that species. Some animals may need to be moved to areas where they are more likely

Fast Fact

Some wildlife biologists work with the public, teaching people about animals and their needs.

to survive. Wildlife biologists may also work to improve a habitat to help a species prosper, such as making sure nesting areas are available for birds.

Some wildlife biologists create human-made islands for waterbirds.

A wildlife biologist sets up a camera to record images of animals in the wild.

A wildlife biologist often has a college degree in wildlife biology, zoology, wildlife ecology, fisheries management, or a similar focus of study. Some have a master's degree or other advanced schooling too. Courses they take include animal behavior, biology, botany, and **statistics**.

Natural Resources Manager

Some scientists work to protect all the organisms and their surroundings—including the soil and water—in an area. They may be called natural resources managers or conservation scientists. Their job duties can be varied but often have to do with **sustainable** land use.

ZOOLOGISTS

Zoologists follow much the same course of study as wildlife biologists. In fact, they may do many of the same jobs. However, zoologists tend to work in zoos, parks, and other controlled environments rather than in the wild. Some may work in rehabilitation facilities, or places that treat hurt animals and release them back into the wild. Zoologists may also focus more on an animal's features and behaviors and less on their habitat. Zoologists and wildlife biologists who specialize in certain creatures may have different names. For example, entomologists study insects, herpetologists study reptiles and amphibians, ichthyologists study fish, and ornithologists study birds.

Jane Goodall is a famous primatologist. That means she studies **primates**.

Fast Fact

Zoologists and wildlife biologists may track animals' movements through GPS and other technology.

An elk in California wears a GPS tracker that will let scientists track its movements.

For example, a natural resources manager may work with a logging company to determine how much of a forest can be cut down for people's use without harming the habitat. They may direct when and where to plant new trees. A natural resources manager may be hired to observe and suggest changes to fishing rules in a bay. They might provide data about plant and animal life in areas where drilling and mining take place. They may also observe the impact of tourists on a national or state park. In fact, many natural resources managers work for government agencies, such as the National Park Service and the U.S. Department of Agriculture.

Fast Fact

A conservation officer is someone who enforces laws about wildlife and nature. They can fine or even arrest people who break the law. They may work with natural resources managers.

Most natural resources managers have a college degree in a science field. However, employers might consider a high school diploma and work experience as a substitute for this.

A natural resources manager may help with land rehabilitation efforts after an event such as a forest fire or flood.

The oceans need to be healthy for the organisms that live in them, and for the health of life on land as well. Tiny ocean creatures called phytoplankton produce more than half the oxygen we need to breathe!

Chapter Four

INTO THE OCEAN

There are so many areas of Earth that need people's attention and that environmental scientists can help protect. The ocean, for instance, covers about 70 percent of the planet's surface. It's home to organisms that we depend on for food, oxygen, and much more. Ocean currents help regulate Earth's climates, making them less extreme. Ocean waters are so vast that they still hold a lot of mysteries for environmental scientists.

Oceanographer

The study of the ocean is called oceanography. Like other environmental scientists, oceanographers usually focus on a smaller area of research. Studying biology, physics, chemistry, and geology in college is an effective way to prepare for an oceanography career. Different kinds of oceanographers often work together on research teams.

> **Fast Fact**
>
> Only 5 percent—or less!—of Earth's oceans have been explored so far.

Oceanographers who study the movement of ocean currents and how they impact weather and climate are called physical oceanographers. Geological oceanographers study

Garbage in the ocean is a growing problem. People are still learning about the harmful effects plastics and other waste have on marine creatures.

the ocean floor and its movements, including underwater earthquakes and volcanoes. Chemical oceanographers study the chemical makeup of water, including the effects of pollution on ocean organisms. Perhaps the most familiar of all oceanographers is the marine biologist.

Fast Fact

Some geological oceanographers study the Mariana Trench, the deepest trench on Earth. It's nearly 7 miles (11.3 km) deep!

Marine Biologist

Marine biologists are oceanographers who study life in saltwater bodies. These organisms range from **microscopic** plankton to the largest organism on Earth, the blue whale. It's common for marine biologists to focus on a species or group of animals or plants, animal behavior, or marine habitat. Marine scientists use technology, including **sonar** and remotely operated vehicles (ROVs), to learn more about what's going on in the deep ocean.

An ROV can stay underwater much longer and go much deeper than human divers can.

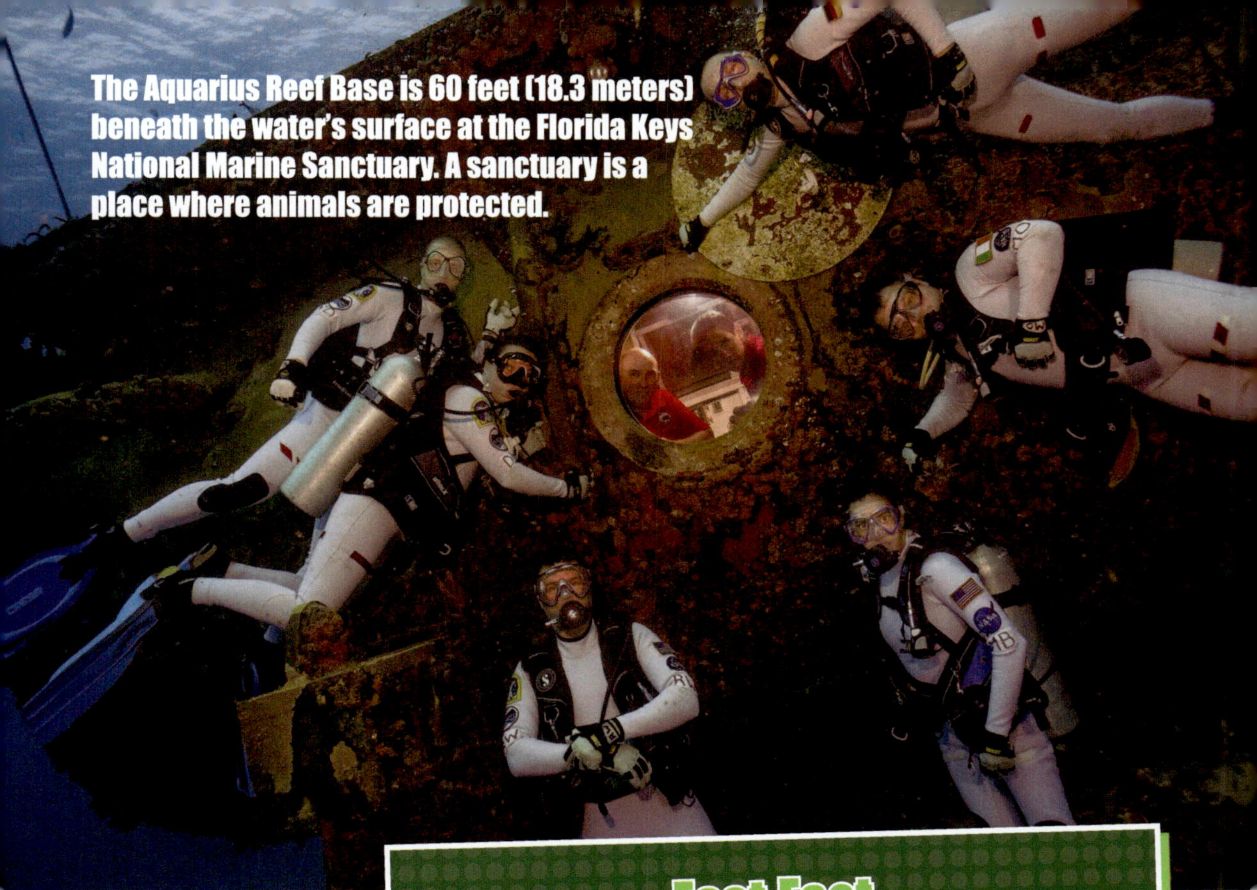

The Aquarius Reef Base is 60 feet (18.3 meters) beneath the water's surface at the Florida Keys National Marine Sanctuary. A sanctuary is a place where animals are protected.

Some marine biologists spend a lot of time on the open ocean tracking animal populations. Others work in coral reefs. Some work at an underwater base called the Aquarius Reef Base, located in the Florida Keys. Still others spend most of their time on land, at an aquarium or school, for example.

Fast Fact

In 2012, marine biologist Edith Widder was the first person to film a giant squid in ocean waters.

Dr. Edith Widder first became interested in oceanography when she visited the island country of Fiji in the South Pacific Ocean at age 11.

BE A TEACHER!

If you can't decide what field to focus on, but know you have a passion for science and the environment, you might want to be an environmental science teacher. You'll cover all kinds of environmental subjects as you prepare to teach and may be able to center your teaching on the topics you care about. You can decide the best ways to excite the next generation of environmental scientists and **activists** about protecting the planet. Environmental science teachers often teach in high school but may teach younger or older students too. They likely have a college degree and a master's degree—or more schooling to teach at a higher level.

Teaching is a way to make a big impact on young people's lives.

Marine biologists often earn a bachelor's degree. If they wish to pursue their own research or lead a team, they may have to get an advanced degree.

Experience Is Essential

No matter their interest, environmental scientists can make a major difference in the world today. They can figure out ways to save plants and animals from destructive human activities. The comeback of the once almost-extinct bald eagle is an example of what can happen when scientists, governments, and many other people take action.

While many of the careers in this book require a college degree, getting experience now is a way to test out a future career. Look for volunteer opportunities in environmental science that interest you. You'll likely get to spend time outdoors, learn something new, and help your community!

Sometimes volunteer opportunities can lead to future careers, and they're also a great way to make friends with people who share your interests. Look for ways to get involved through a local wildlife or conservation organization. You can find volunteer opportunities online!

MORE ENVIRONMENTAL SCIENCE JOBS

hydrologist
studies how water moves on the surface and under the ground

land use planner
works with others to decide the best use of land within a community with the least impact on the environment

agricultural scientist
studies how crops and animals can be raised for food in more sustainable ways

science editor
prepares environmental research to be read and understood by the public

ecologist
studies the relationship between plants, animals, and other organisms and their habitats

microbiologist
studies organisms such as bacteria, viruses, algae, fungi, parasites, and other things too small to be seen with the human eye

forest ranger
educates the public about natural resources, enforces laws, and helps people and animals in need

industrial ecologist
studies the effects of industries on the environment and how to reduce harmful impacts

THINK ABOUT IT!

1. Why do you think more environmental scientists are expected to be needed in the future?
2. Think of an example where different kinds of environmental scientists might work together.
3. Do you think environmental science should be a more significant part of school for younger students? Why or why not?
4. Of all the environmental science jobs described in this book, which do you find most interesting, and why?

GLOSSARY

activist: A person who uses or supports strong actions to help make changes in politics or society.

analyst: A person whose job is to analyze something, or study it closely, in order to discover something about it.

asthma: A long-term condition that affects the airways in the lungs and sometimes causes difficulties breathing.

contiguous: Sharing borders.

GPS: Stands for Global Positioning System. A system that uses satellite signals to locate places on Earth.

habitat: The natural place where an animal or plant lives.

jurisdiction: The area over which laws can be enforced.

microscopic: Having to do with very small objects that can only be seen using a tool called a microscope that makes them look much larger.

primate: Any animal from the group that includes humans, apes, and monkeys.

sonar: A method or machine that uses sound waves to find things in a body of water.

statistics: A branch of mathematics that involves collecting and studying numerical data.

sustainable: Describing doing something using methods that don't use up or destroy natural resources.

technology: A method that uses science to solve problems and the tools used to solve those problems.

FIND OUT MORE

Books

Adams, William D. *Environmental Chemistry*. Chicago, IL: World Book, 2023.

Gagne, Tammy. *Environmental Engineer*. New York, NY: AV2, 2021.

Morkes, Andrew, and Shaina C. Indovino. *Women in Environmental Sciences*. Hollywood, FL: Mason Crest, 2022.

Websites

The Environment
www.ducksters.com/science/environment/
Read more about the environment itself and issues relating to it.

So, You Want to Be a Wildlife Biologist?
www.maine.gov/ifw/blogs/mdifw-blog/so-you-want-be-wildlife-biologist
Read more about what wildlife biologists do.

What Does a Marine Biologist Do?
www.wonderopolis.org/wonder/what-does-a-marine-biologist-do
Find out more about this environmental science career involving ocean life.

INDEX